Prentice Hall Writing and Grammar

Communication in Action

Silver Level

Writing Support Activity Book

Prentice Hall

Upper Saddle River, New Jersey
Glenview, Illinois
Needham, Massachusetts

ISBN 0-13-043893-6

6 7 8 9 10 04

Prentice Hall

Contents

Introduction

This workbook contains graphic organizers designed to help you through the writing process. The organizers correspond to the writing strategies presented in your *Writing and Grammar* textbook. After you learn about a particular strategy in your textbook, you can apply it to your own work. Simply turn to the corresponding graphic organizer and use it to help you gather notes and ideas for your writing.

The last section of the workbook contains ten additional, generic graphic organizers, which you can use to plan your writing or to organize information from your reading.

Name:______________________________ Date:__________________

Chapter 1: The Writer in You

Writing Support Activity 1–1: Writing Calendar

Month:

Sunday	Monday	Tuesday	Wednesday	Thursday	Friday	Saturday

Name:______________________________ Date:____________________

Chapter 1: The Writer in You

Writing Support Activity 1–2: Technology Evaluation Chart

TOPICS	 Writing Tools	 Virtual Resources	 Audiovisual Tools
Exercise			
Pet Care			
Study Skills			

Name:________________________________ Date:____________________

Chapter 2: A Walk Through the Writing Process

Writing Support Activity 2–1: Topic Web

Broad Topic: ______________________

Broad Topic: ______________________

Subtopic: __________

Subtopic: __________

Subtopic: __________

Narrowed topic: __________________________

Name:________________________________ Date:__________________

Chapter 2: A Walk Through the Writing Process

Writing Support Activity 2–2: Audience Profile

AUDIENCE PROFILE

- What is the average age of my audience?

- What do they know about my topic?

- What details will be most interesting to my audience?

- What background do I need to provide?

Name:______________________________ Date:______________________

Chapter 2: A Walk Through the Writing Process

Writing Support Activity 2–3: Parts-of-Speech Word Web

NOUNS:

VERBS:

TOPIC:

____

ADJECTIVES:

ADVERBS:

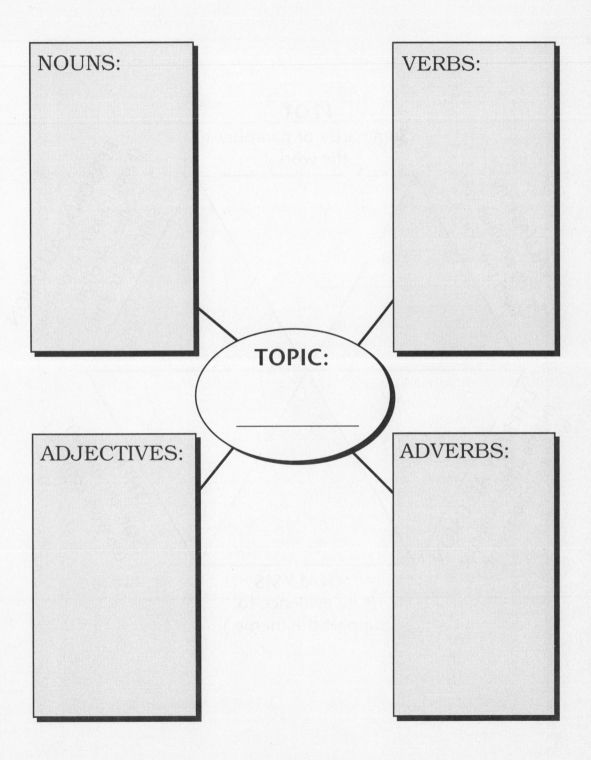

Name:______________________________ Date:__________________

Chapter 2: A Walk Through the Writing Process

Writing Support Activity 2–4: Hexagon

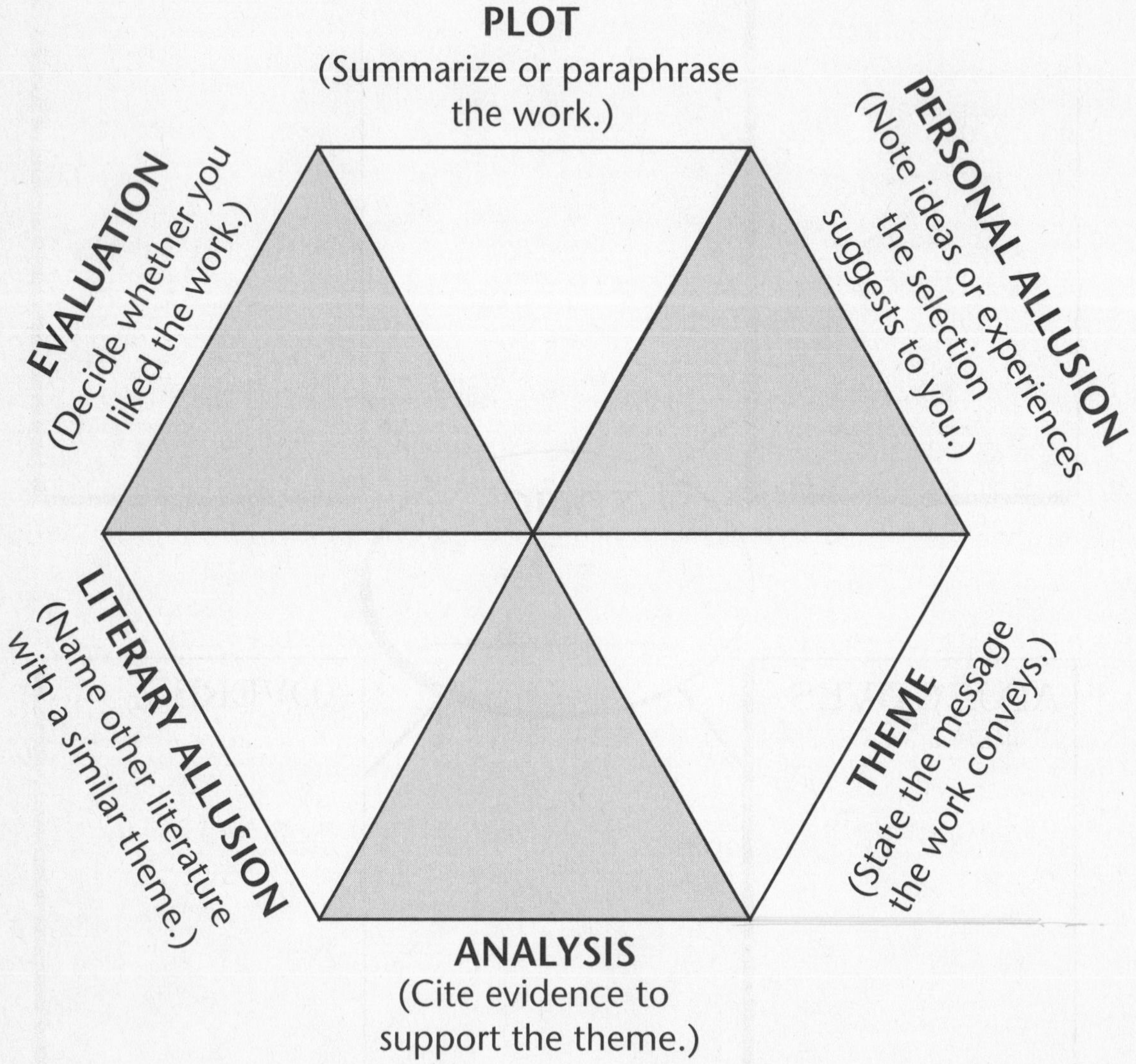

PLOT
(Summarize or paraphrase the work.)
PERSONAL ALLUSION
(Note ideas or experiences the selection suggests to you.)
THEME
(State the message the work conveys.)
ANALYSIS
(Cite evidence to support the theme.)
LITERARY ALLUSION
(Name other literature with a similar theme.)
EVALUATION
(Decide whether you liked the work.)

Name:________________ Date:__________

Chapter 2: A Walk Through the Writing Process

Writing Support Activity 2–5: Question-and-Answer Outline

Topic: ________________

Question 1:

- •
- •
- •
- •
- •
- •

Question 2:

- •
- •
- •
- •

Name:______________________________ Date:____________________

Chapter 3: Paragraphs and Compositions: Structure and Style

Writing Support Activity 3–1: Media Comparison Venn Diagram

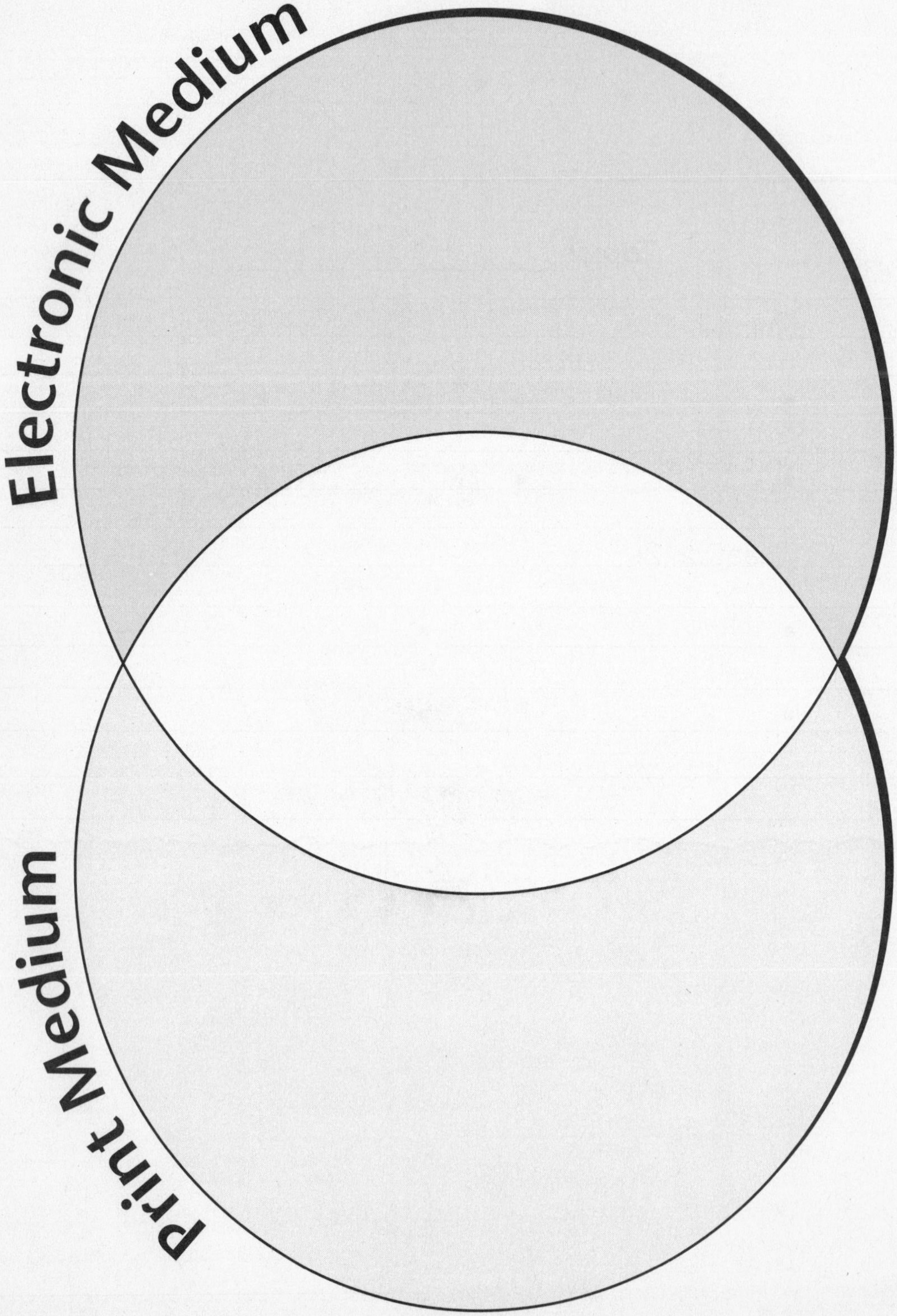

Name:______________________________ Date:____________________

Chapter 4: Narration: Autobiographical Writing

Writing Support Activity 4–1: Emphasizing Tension

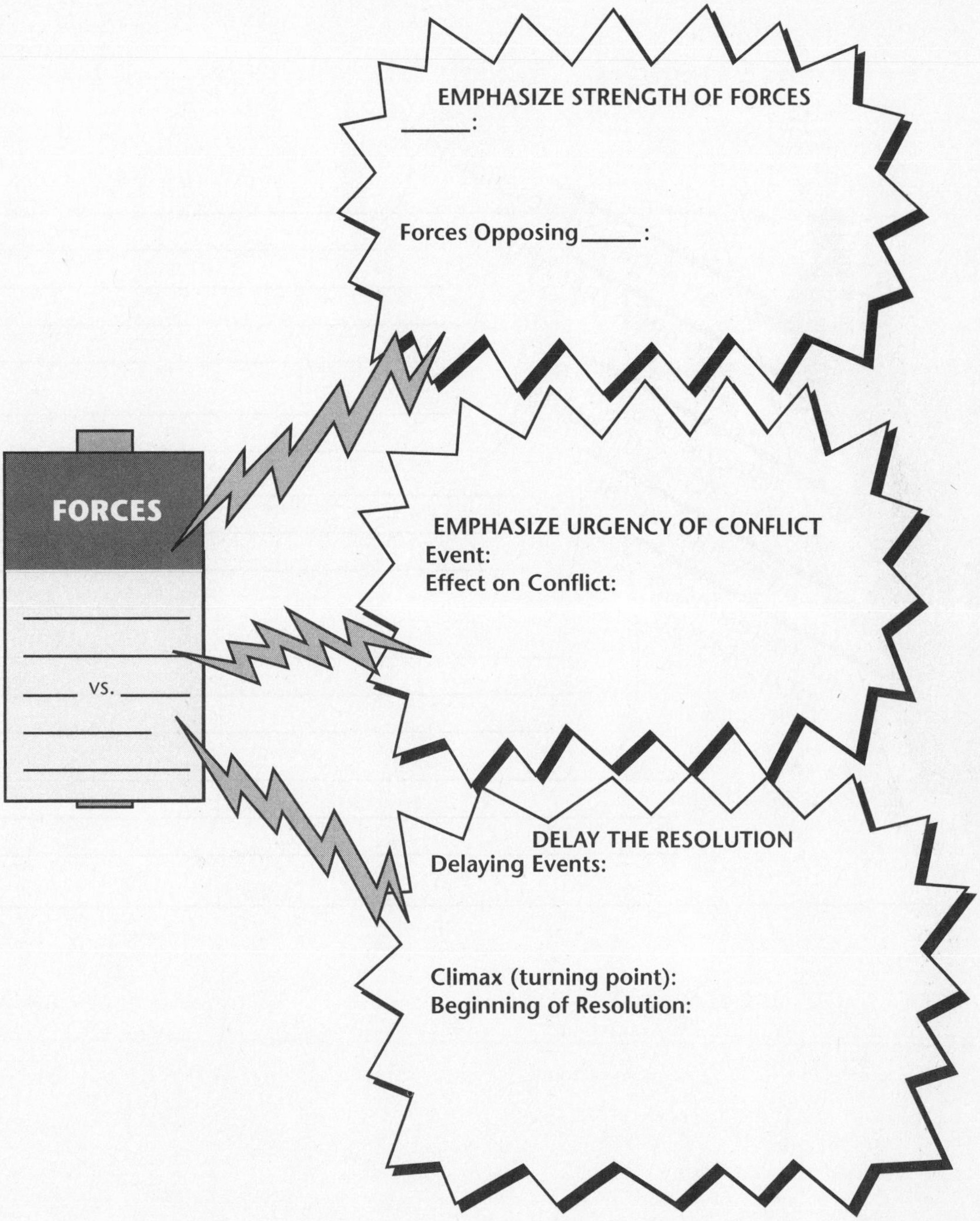

Name:________________________________ Date:________________

Chapter 4: Narration: Autobiographical Writing

Writing Support Activity 4–2: Creating a Good Hook

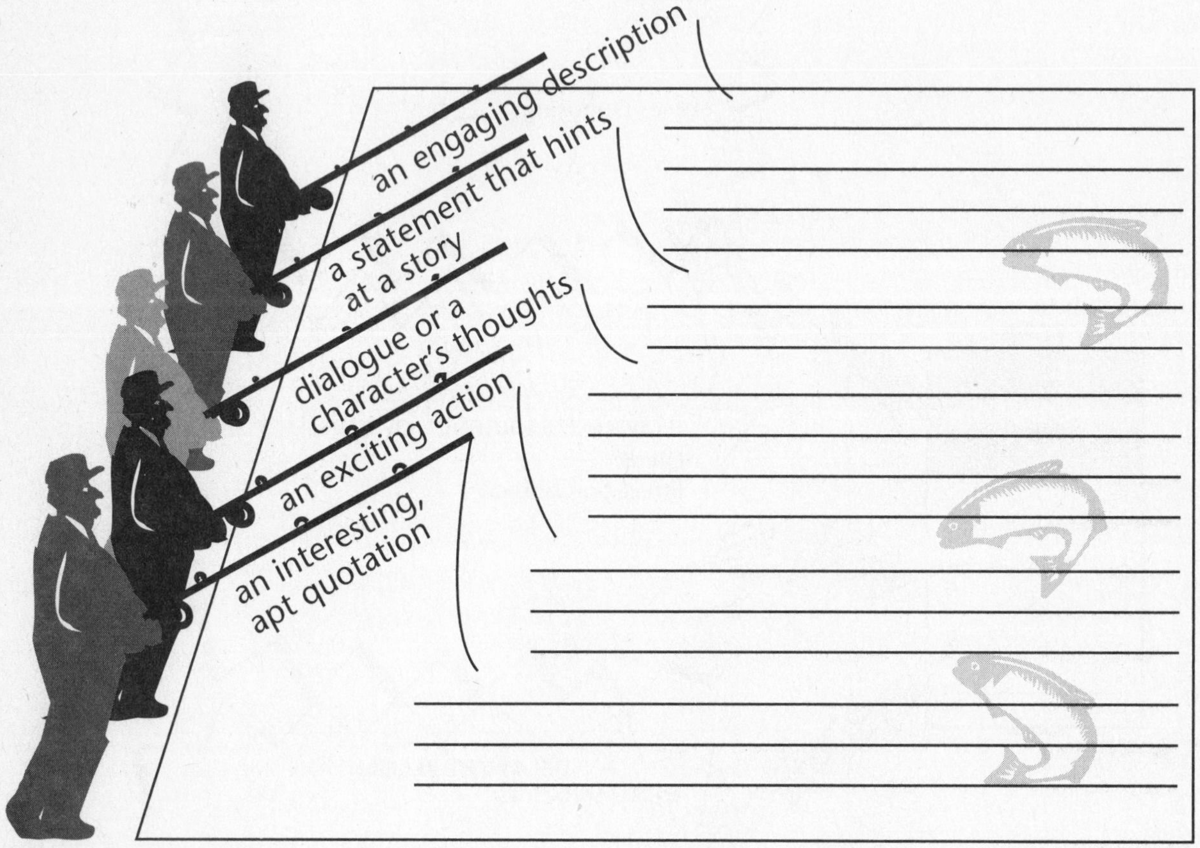

Name:____________________________ Date:____________________

Chapter 4: Narration: Autobiographical Writing

Writing Support Activity 4–3: Firsthand Biography Chart

Person's Name: ____________________

Last Fall	
Early Spring	
Late Spring	
Summer	
This Fall	

Name:________________________________ Date:________________

Chapter 4: Narration: Autobiographical Writing

Writing Support Activity 4–4: Photo Essay Chart

Aspect of Myself	Type of Photograph (Portrait, Action Shot, Still Life, Landscape/Interior)	Special Considerations

Name:________________________________ Date:____________________

Chapter 5: Narration: Short Story

Writing Support Activity 5-1: Plot Diagram

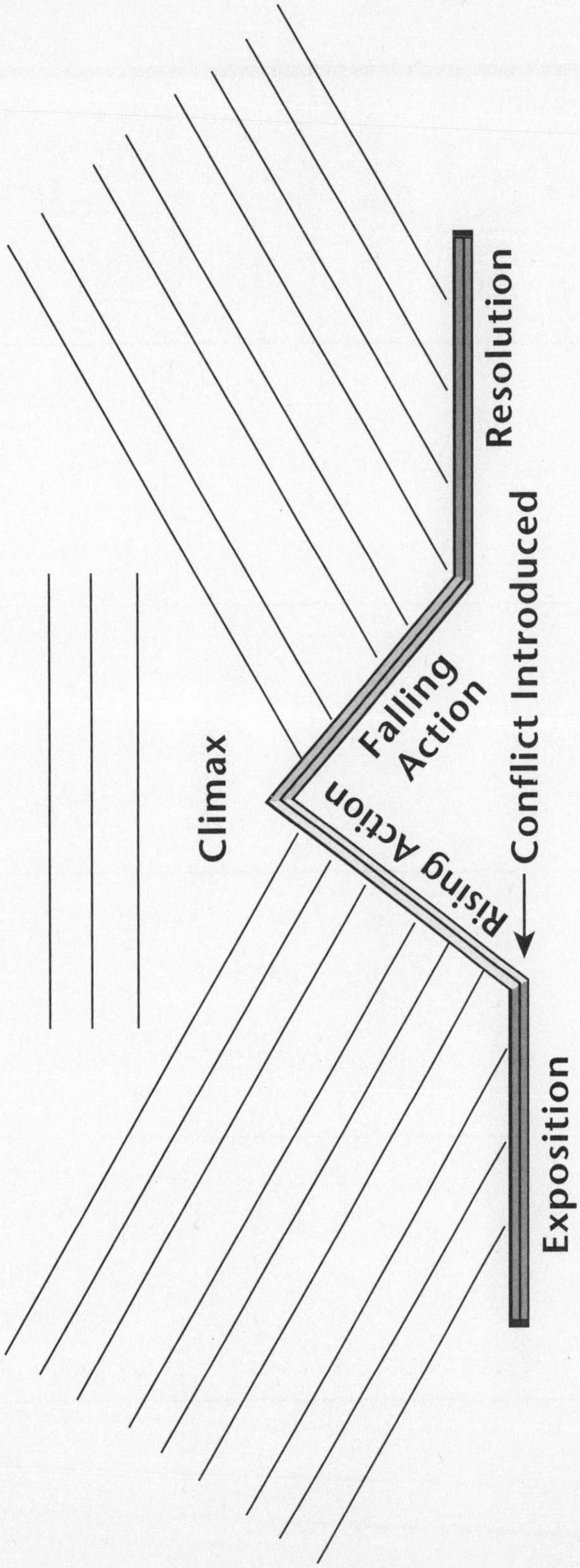

Name:______________________________ Date:____________________

Chapter 5: Narration: Short Story

Writing Support Activity 5–2: Characterization Chart

Character's Name: ____________________

Background	Style of Speaking	Situation	Effect of Background on Situation	Attitude Toward Situation

Name:______________________________ Date:__________________

Chapter 5: Narration: Short Story

Writing Support Activity 5-3: Analyzing Cartoons

Character: Distinctive Physical/Personality Characteristics	Conflict	Events to Which Conflict Leads	Theme (question or message about life) Expressed by Conflict and Events

Name:________________________________ Date:________________

Chapter 6: Description

Writing Support Activity 6–1: Cubing

How does it look, sound, smell, feel, or taste?

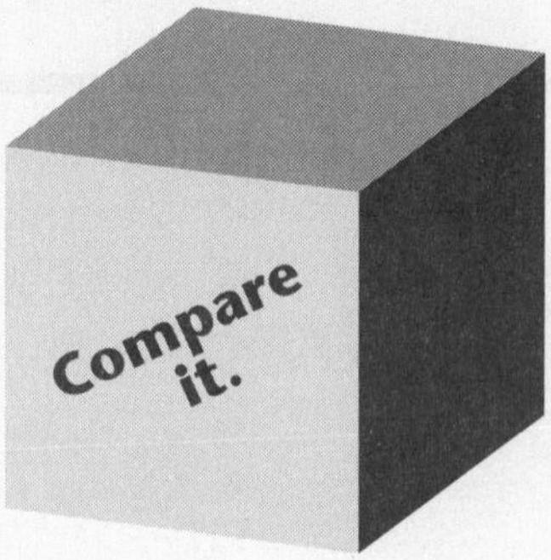

What is it similar to? Different from?

What does it make you think of?

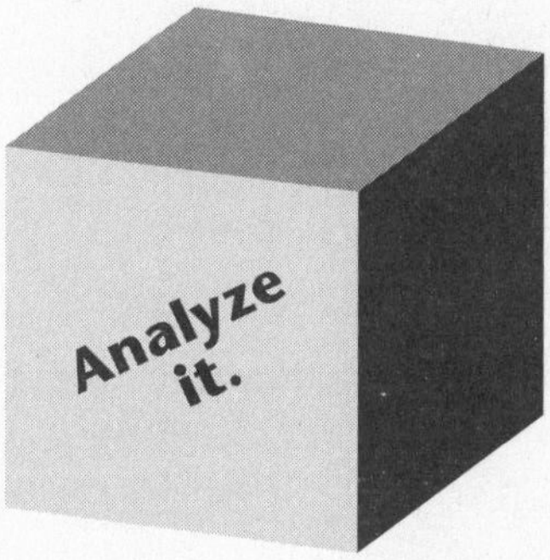

How is it made? How does it work?

How does it fit into your experience? How can you use it?

Is it a positive or a negative? Helpful or harmful?

Name:______________________________ Date:__________________

Chapter 6: Description

Writing Support Activity 6–2: Double Dyad

	Very Well	Okay	Not Very Well
How well does the introduction capture your attention?			
How well does the writer use sensory details and comparisons to create a picture of the subject?			
Does the essay flow?			
Does the writer create a main impression of the subject?			
Does the conclusion leave a memorable impression?			

What is the strongest part of the paper? ______________________________

____

____

____

____

What is the weakest part of the paper? ______________________________

____

____

____

____

Name:______________________________ Date:________________

Chapter 6: Description

Writing Support Activity 6–3: Poem Topic Web

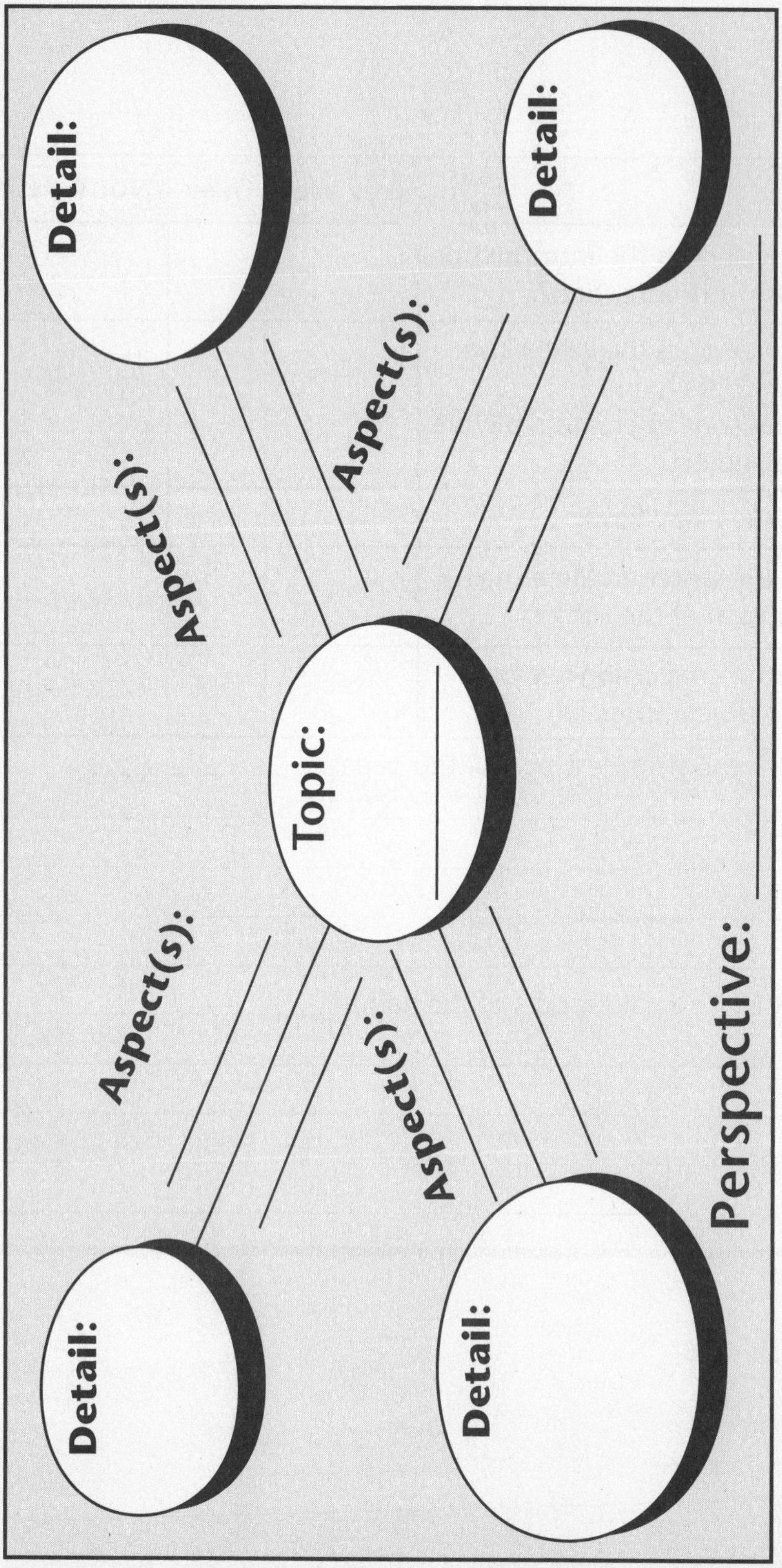

Name:______________________________ Date:____________________

Chapter 6: Description

Writing Support Activity 6–4: Analyzing Camera Techniques

Movie: ______________________________

Action	Technique Used (Editing, Camera Movement, Camera Angle, Framing, etc.)	Effect of Technique	Role of Viewer (Follows Action/ Is Above Action/ Misses Part of Action/Shocked by Action, etc.)

Name:______________________________ Date:____________________

Chapter 7: Persuasion: Persuasive Essay

Writing Support Activity 7-1: T-Chart

Issue: ____________________

Pro	Con

Name:____ Date:______________________

Chapter 7: Persuasion: Persuasive Essay

Writing Support Activity 7–2: Double Dyad

1. Exchange papers with a partner.
2. Use the following grid to rate your partner's paper.
3. Discuss the responses with your partner.
4. Repeat the process with another partner.

	Very Well	Okay	Not Well
How well does the introduction create interest?			
How well does the writer present the main argument?			
Does the essay flow?			
Does the writer prove his or her argument?			
Does the conclusion wrap up the argument?			

What is the strongest part of the paper?

____

____

____

What is the weakest part of the paper?

____

____

____

Name:______________________________ Date:________________

Chapter 7: Persuasion: Persuasive Essay

Writing Support Activity 7–3: Advertisement Planner

Television Commercial for ______________________

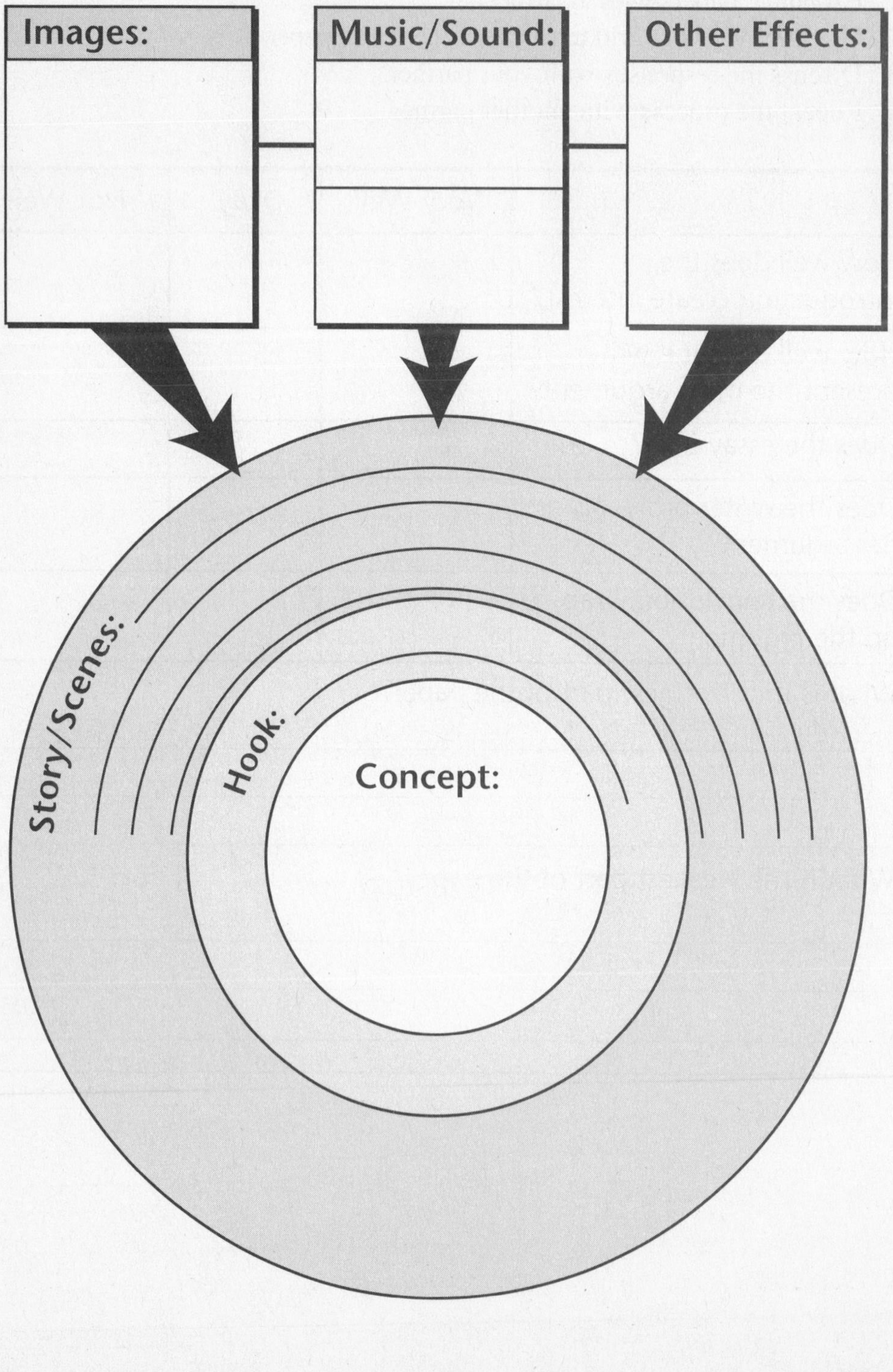

Name: ____________________ Date: ____________________

Chapter 7: Persuasion: Persuasive Essay

Writing Support Activity 7–4: Analyzing Commercials

Commercial for: ____________________

Summary of Action	Description of Actors/ Speakers	General Theme ("Fun," "Power," and so on)	Specific Message	Techniques Used to Create Message

Name:________________________________ Date:________________

Chapter 8: Exposition: Comparison-and-Contrast Essay

Writing Support Activity 8–1: Personal-Experience Timeline

Use this timeline to chart ways you have changed over time.
Choose two entries as the basis for a comparison-and-contrast essay.

Topic: ____________________

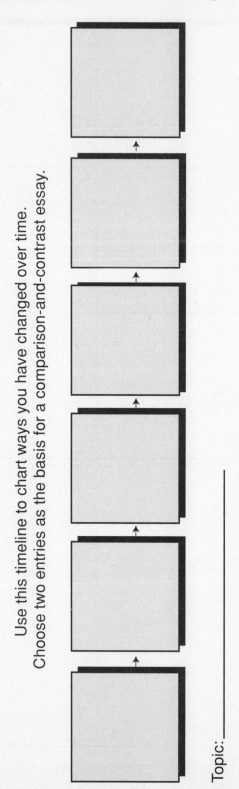

Name:______________________________ Date:______________

Chapter 8: Exposition: Comparison-and-Contrast Essay

Writing Support Activity 8–2: Venn Diagram

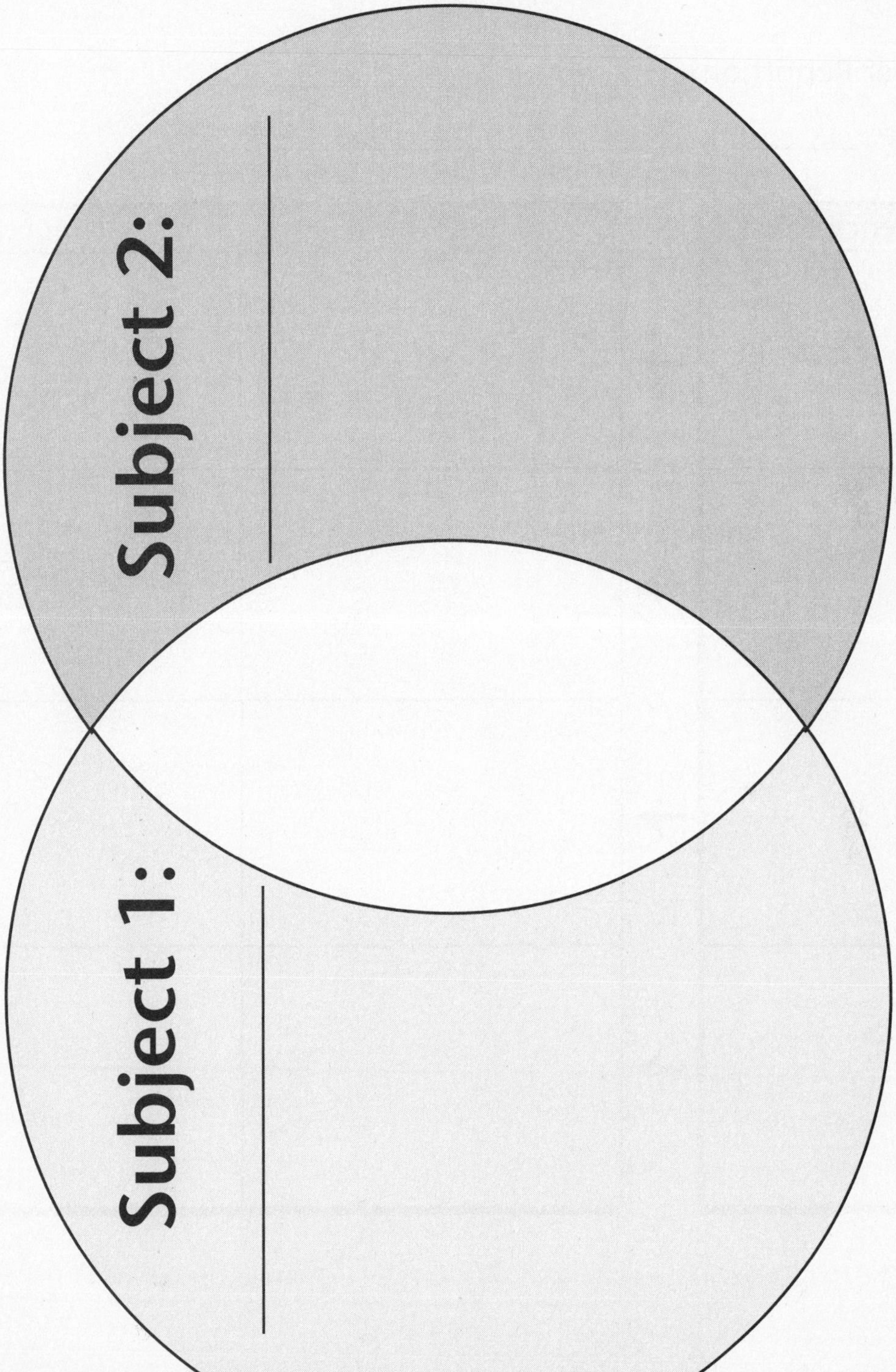

Name:________________________________ Date:__________________

Chapter 8: Exposition: Comparison-and-Contrast Essay

Writing Support Activity 8–3: Consumer Report Chart

Consumer Report on: ______________________

Audience: ______________________________

Audience Need		Product 1	Product 2
	→		
	→		
	→		
	→		

Name:________________________________ Date:____________________

Chapter 8: Exposition: Comparison-and-Contrast Essay

Writing Support Activity 8–4: Book and Movie Comparison Chart

Title: ______________________________

	Written Version	Movie Version	Possible Reason for Difference	Does Difference Improve Movie?
Events				
Characters				
Special Effects (Written Descriptions and Figurative Language vs. Lighting, Color, Music, Camerawork)				

Name:____________________ Date:____________

Chapter 9: Exposition: Cause-and-Effect Essay

Writing Support Activity 9–1: Self-Interview

Questions	Answers
1.	
2.	
3.	
4.	
5.	
6.	
7.	
8.	
9.	
10.	

Name:____________________ Date:______________

Chapter 9: Exposition: Cause-and-Effect Essay

Writing Support Activity 9–2: Classical Invention

General Topic: ____________________

- In what group or general category does your topic belong?

- How is your topic similar to or different from other topics in this category?

Narrowed Topic

____

____

____

____

____

- What causes are involved with this topic?

- What effects are involved in this topic?

- What came before this event?

- What might come (or came) after this event?

Name:______________________________ Date:____________________

Chapter 9: Exposition: Cause-and-Effect Essay

Writing Support Activity 9–3: K-W-L Chart

What I **L**earned	
What I **W**ant to Know	
What I **K**now	

Name:________________________ Date:____________

Chapter 9: Exposition: Cause-and-Effect Essay

Writing Support Activity 9–4: Documentary Video Script

Title of Documentary: ______________________

SHOT ____		SHOT ____	
Visual:		Visual:	
Camera Angle/ Movement:	→	Camera Angle/ Movement:	→
Narration:		Narration:	
Other Sound:		Other Sound:	

Name:____ Date:____________________

Chapter 10: Exposition: How-to Essay

Writing Support Activity 10–1: Listing

After you have made your list, circle words and draw lines to show connections between items on the lists. These links may suggest a topic.

Things	Activities	Places

Topic: ____________________________

Name:______________________________ Date:____________________

Chapter 10: Exposition: How-to Essay

Writing Support Activity 10–2: Sticky-Note Timeline

Write each step on a sticky note. If you have forgotten a step or two, use the additional sticky notes. Then, draw an arrow to show where the additional step or steps should go.

Name:____________________ Date:____________________

Chapter 10: Exposition: How-to Essay

Writing Support Activity 10–3: Problem-Solution Chart

Problem:____________________

Causes of Problem →	Solution →	Effect on Problem

Name:______________________________ Date:__________________

Chapter 10: Exposition: How-to Essay

Writing Support Activity 10–4: On-line Help Chart

Program: ________________________

My Question	Form of Help (Balloon, on-line manual, manufacturer's Web site, printed manual, "Read Me" file)	Found Answer?	How Easy/Hard to Find? Explain

Name:______________________________ Date:____________________

Chapter 11: Research: Research Report

Writing Support Activity 11–1: Self-Interview

Answer the questions shown. Then, circle words and draw lines to show connections between items on your list. Choose a topic from among these linked items.

People	Places	Things	Events
What interesting people do I know or know about?	What interesting places have I been to or heard about?	What interesting things do I know about?	What interesting events have happened to me or have I heard about?

Name:______________________________ Date:__________________

Chapter 11: Research: Research Report

Writing Support Activity 11–2: Classical Invention

General Topic: ______________________________

- In what general category does your topic belong?

- How is your topic similar to or different from other topics in this category?

Narrowed Topic

____

____

____

____

____

- Into what other topics can your topic be divided?

- What causes and effects does your topic involve?

Name:____________________ Date:________________

Chapter 11: Research: Research Report

Writing Support Activity 11–3: K-W-L Chart

Topic: ________________

Know	Want to Know	Learned

Name:______________________________ Date:______________

Chapter 11: Research: Research Report

Writing Support Activity 11-4: Evaluating On-line Sources

Research Question: ______________________

Name and Address of Web Site	Last Updated?	Sponsor	How Credible Is the Sponsor? Why?	Sources Cited?	Does a Print Source Confirm the Information?

Name:______________________________ Date:____________________

Chapter 12: Response to Literature

Writing Support Activity 12–1: Browsing

STORY	AUTHOR	SUBJECT	CHARACTERS

Name:______________________________ Date:____________________

Chapter 12: Response to Literature

Writing Support Activity 12–2: Pentad

Actors	Who performed the action?
Acts	What was done?
Scenes	When or where was it done?
Agencies	How was it done?
Purposes	Why was it done?

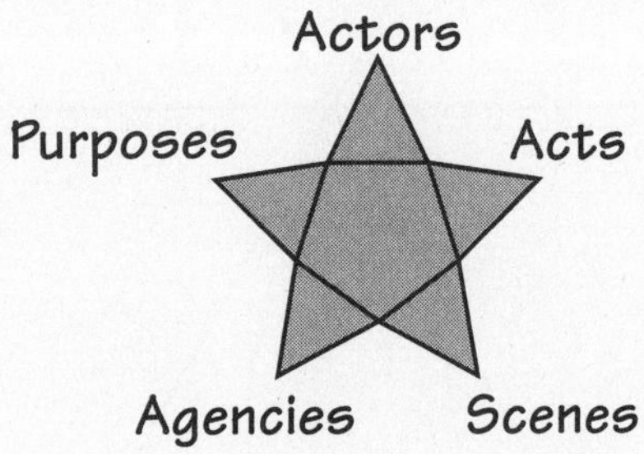

Focus: ______________________________

Name:________________________ Date:________________

Chapter 12: Response to Literature

Writing Support Activity 12–3: Hexagon

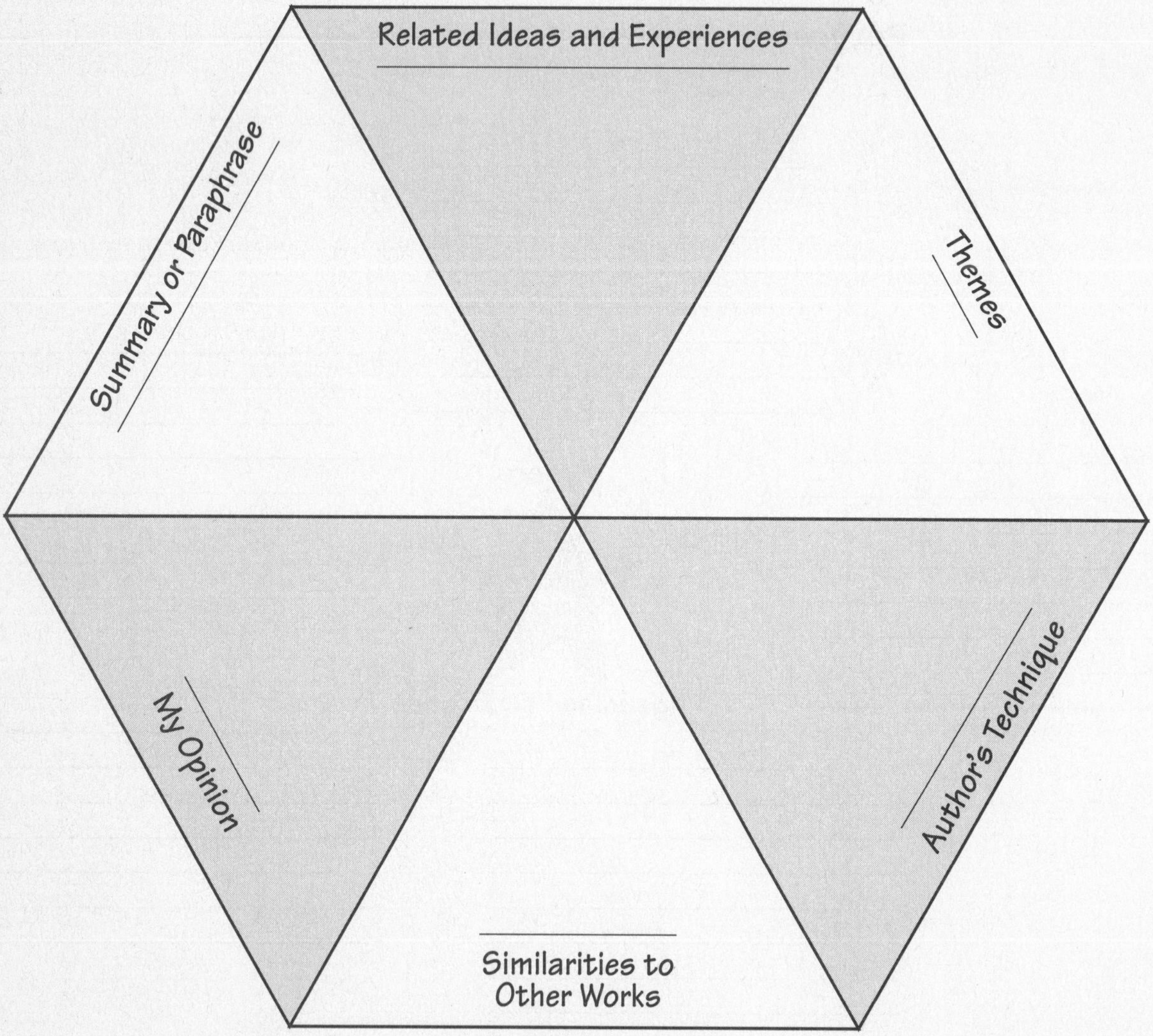

Name:______________________________ Date:__________________

Chapter 12: Response to Literature
Writing Support Activity 12-4: Movie Review Topic Web

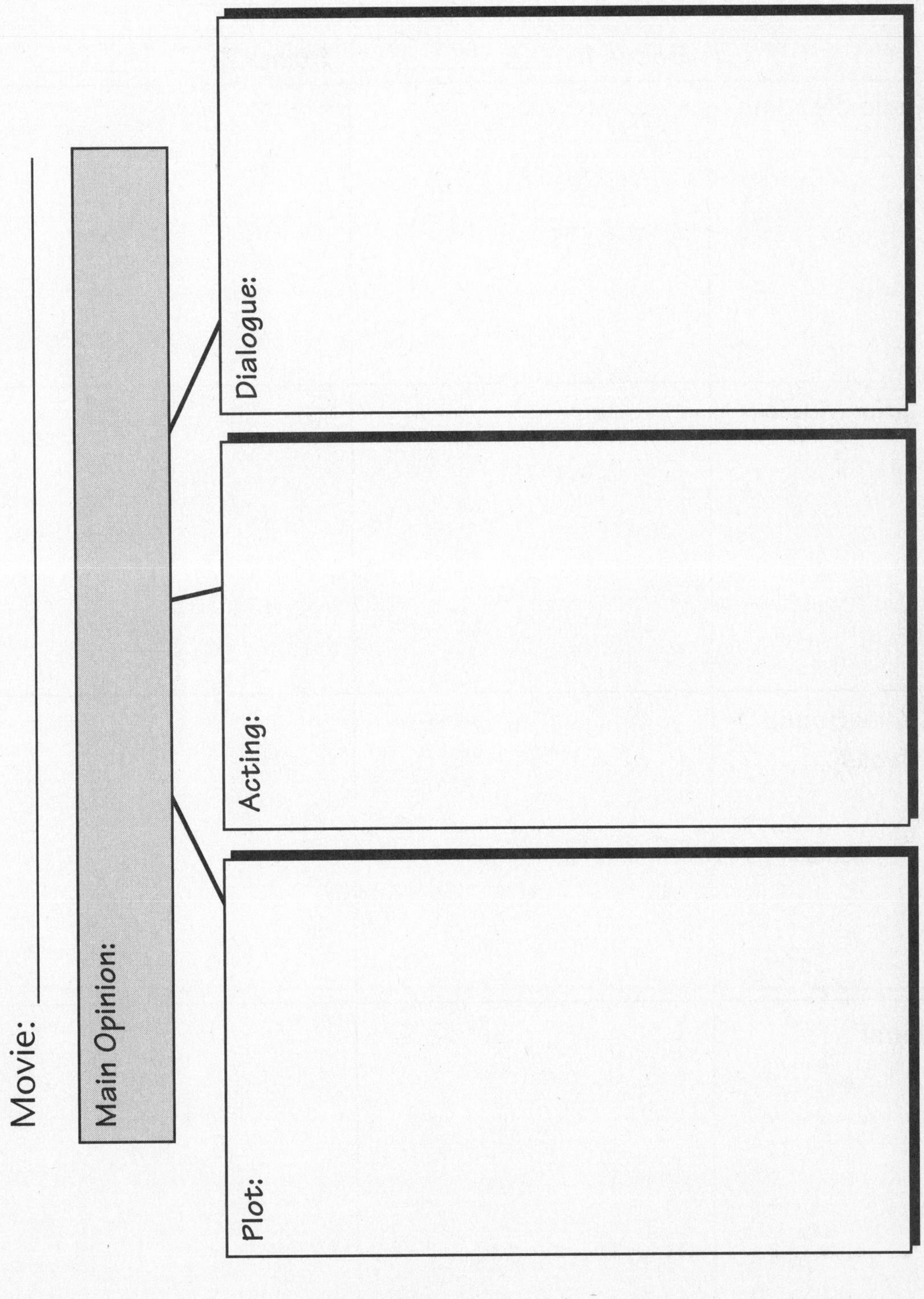

Name:______________________________ Date:______________

Chapter 12: Response to Literature

Writing Support Activity 12–5: Multimedia-Presentation Chart

Presentation of: ______________________

	Scene 1: __________	Scene 2: __________
Audio: Reading		
Audio: Music		
Audio: Sound Effects		
Visual		

Name:______________________________ Date:____________________

Chapter 13: Writing for Assessment

Writing Support Activity 13–1: Timeline

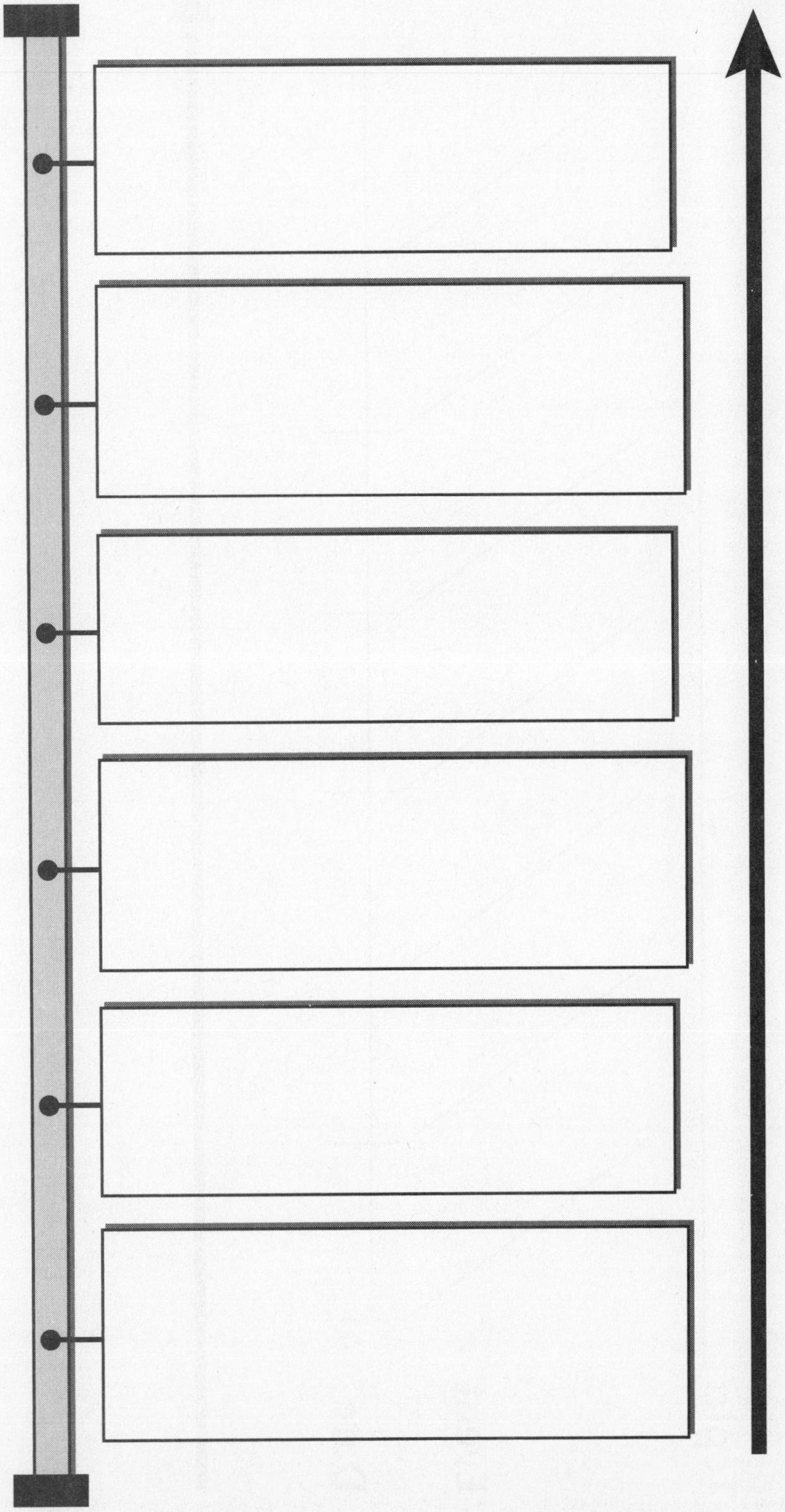

Name:______________________________ Date:________________

Chapter 13: Writing for Assessment

Writing Support Activity 13–2: Open-Book Test Timeline

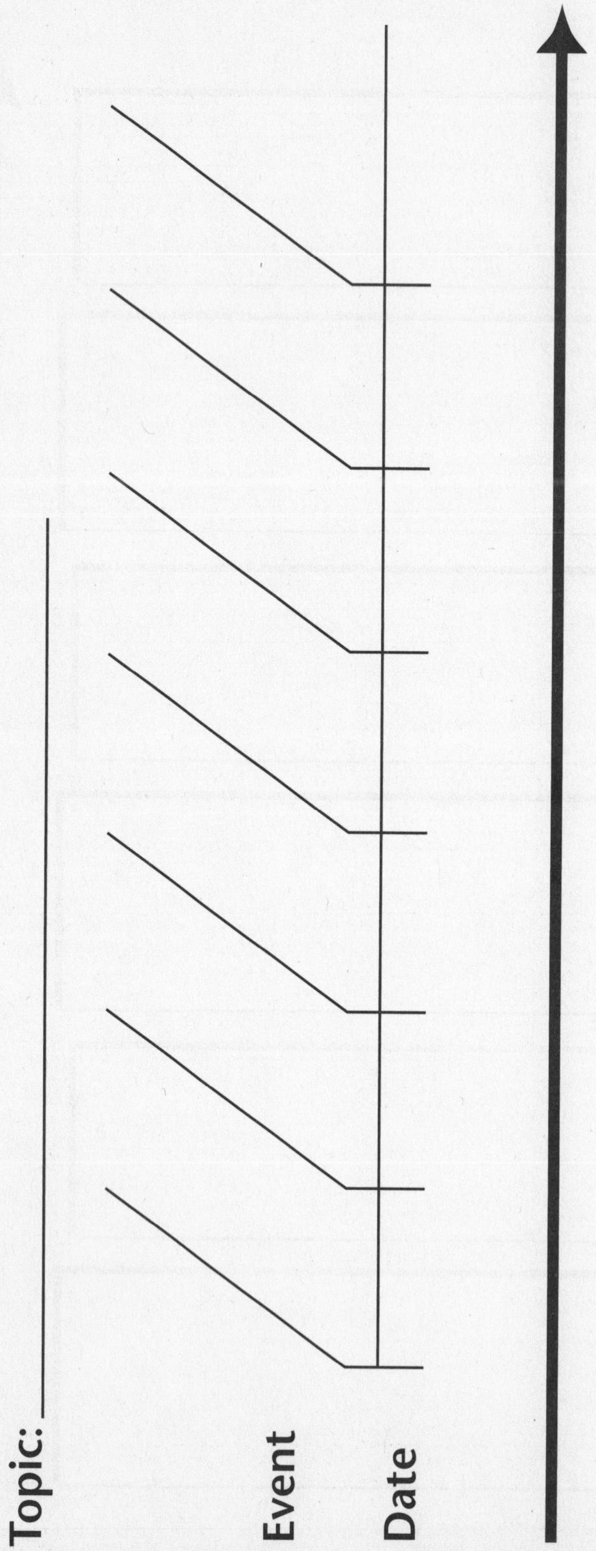

Name:______________________________ Date:__________________

Chapter 13: Writing for Assessment

Writing Support Activity 13–3: Test Analysis Chart

Name of Test: ______________________

How to Answer a Question	How to Move Between Screens	Special Instructions/ Features	Special Problems

Name:______________________ Date:______________

Venn Diagram

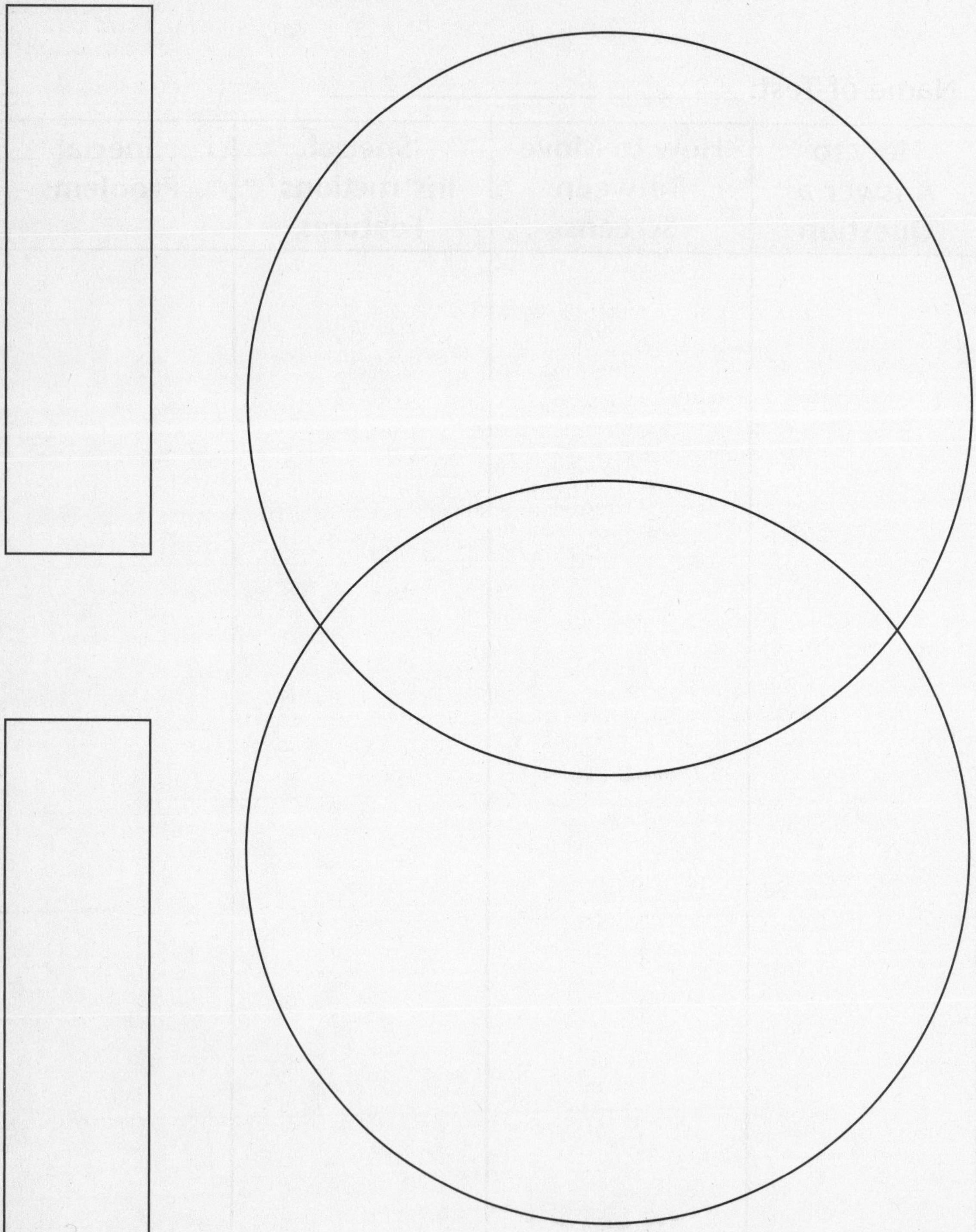

Name:______________________ Date:______________

Comparison and Contrast

Item 1 ______________

Item 2 ______________

Similarities

Differences

Points of Contrast

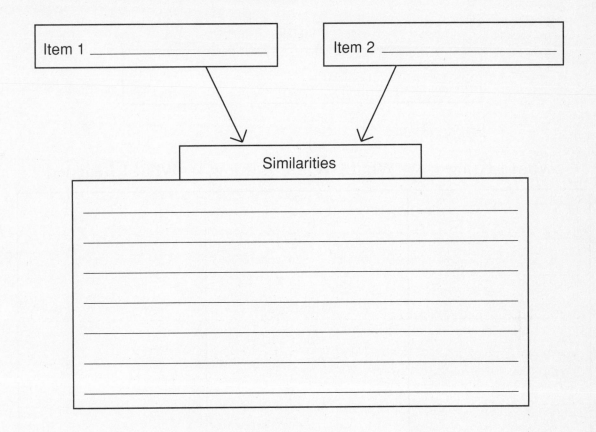

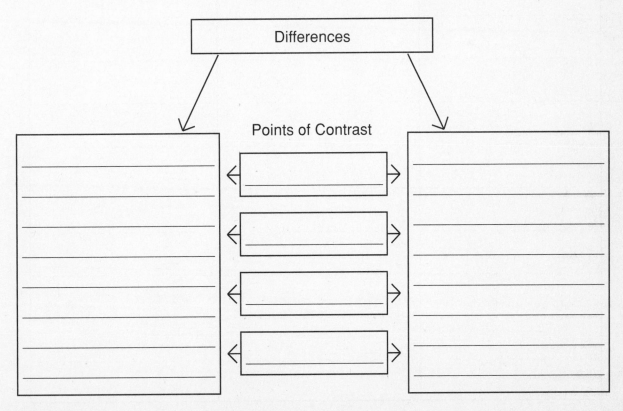

Name:______________________ Date:______________

K-W-L

Topic

What I Know	What I Want to Know	What I Learned

Name:______________________________ Date:__________________

Problem/Solution

Problem	Goal(s)

Alternatives	Pros (+) and Cons (–)
	+
	–
	+
	–
	+
	–
	+
	–

Decision(s)	Reason(s)

Name:________________________ Date:________________

Main Idea and Supporting Details

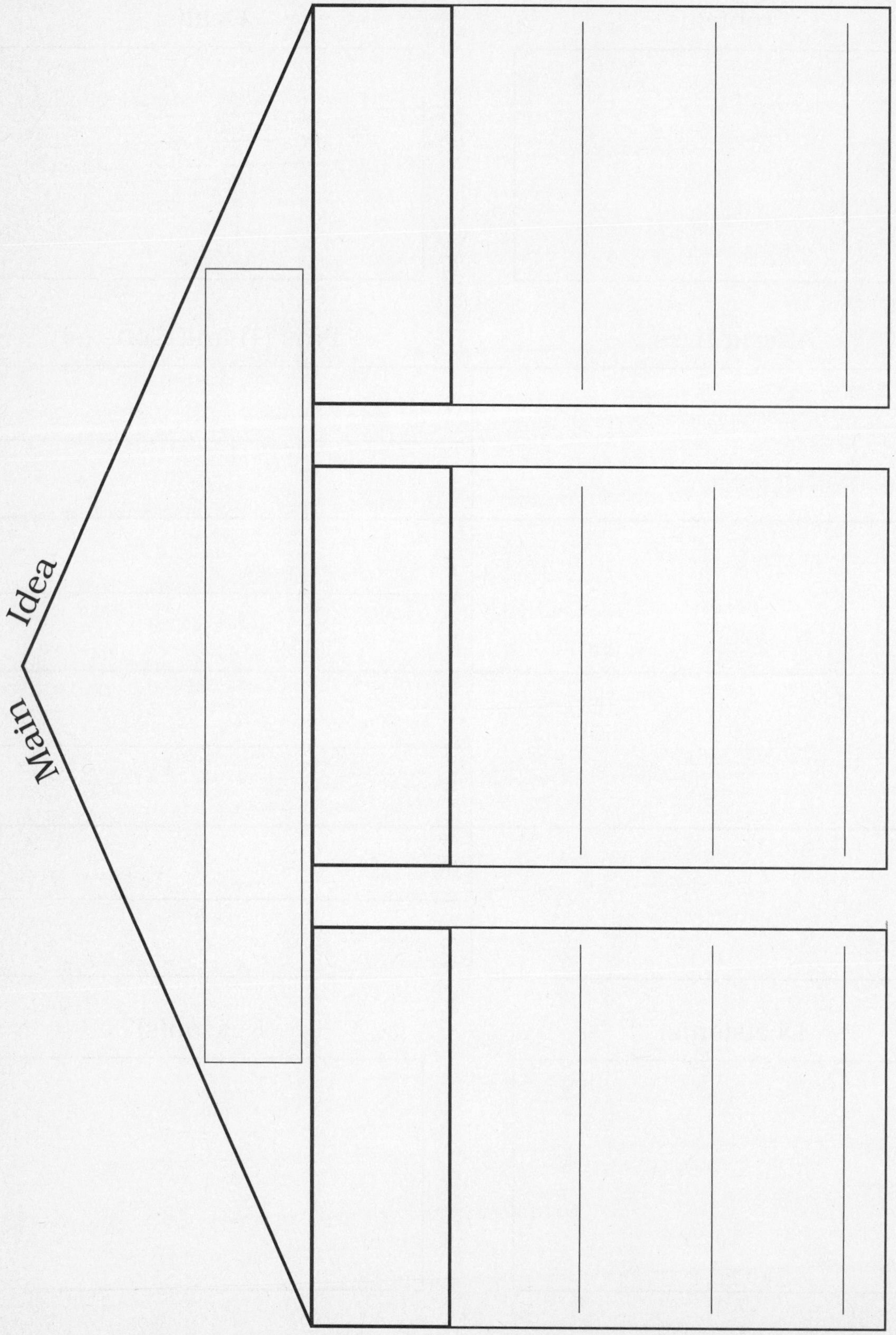

Name:______________________________________ Date:____________________

Two-Column Notes

Topic: ____

Main Idea	Detail Notes

Name:____________________ Date:____________________

Cause and Effect

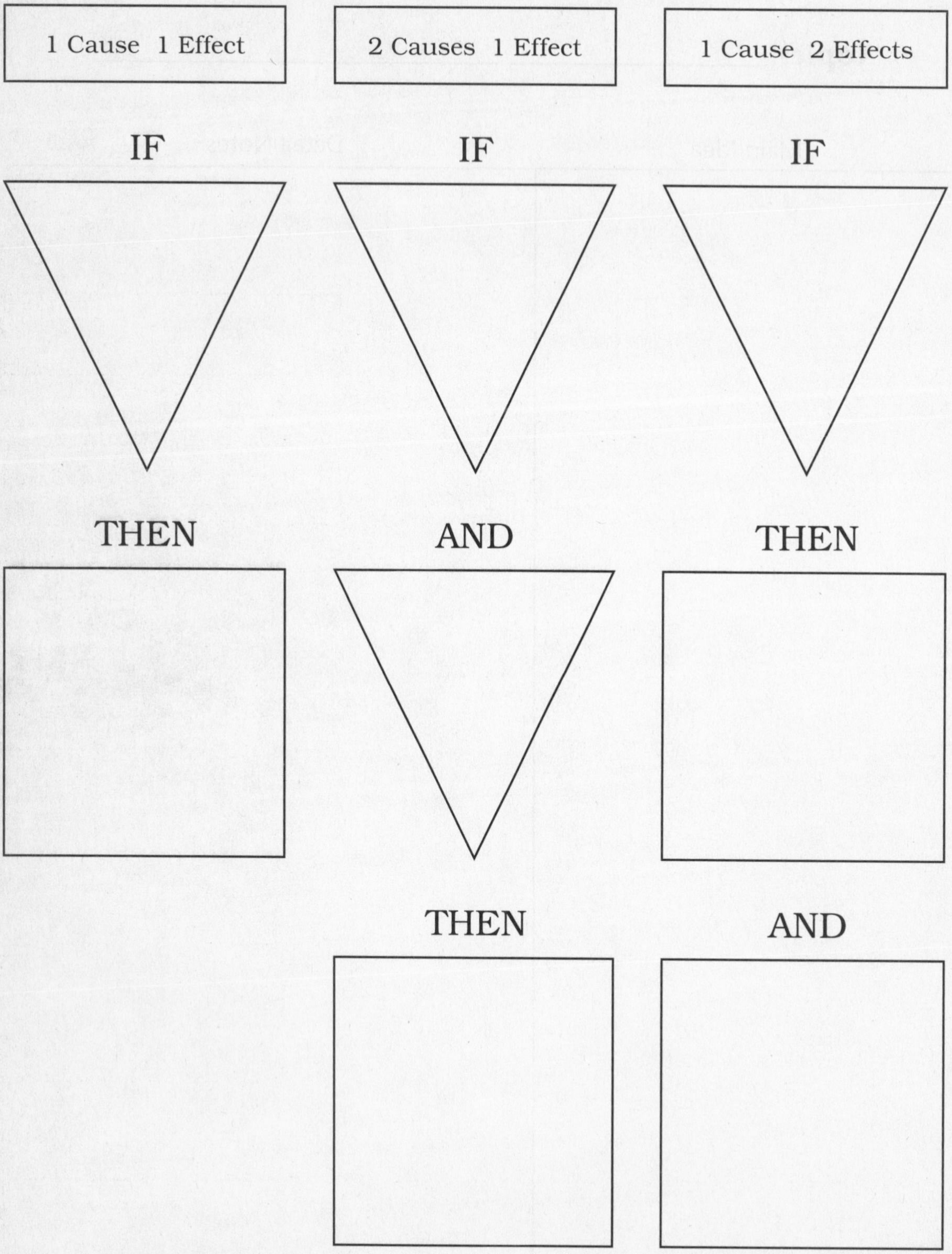

Name:____________________ Date:____________________

Concept Map

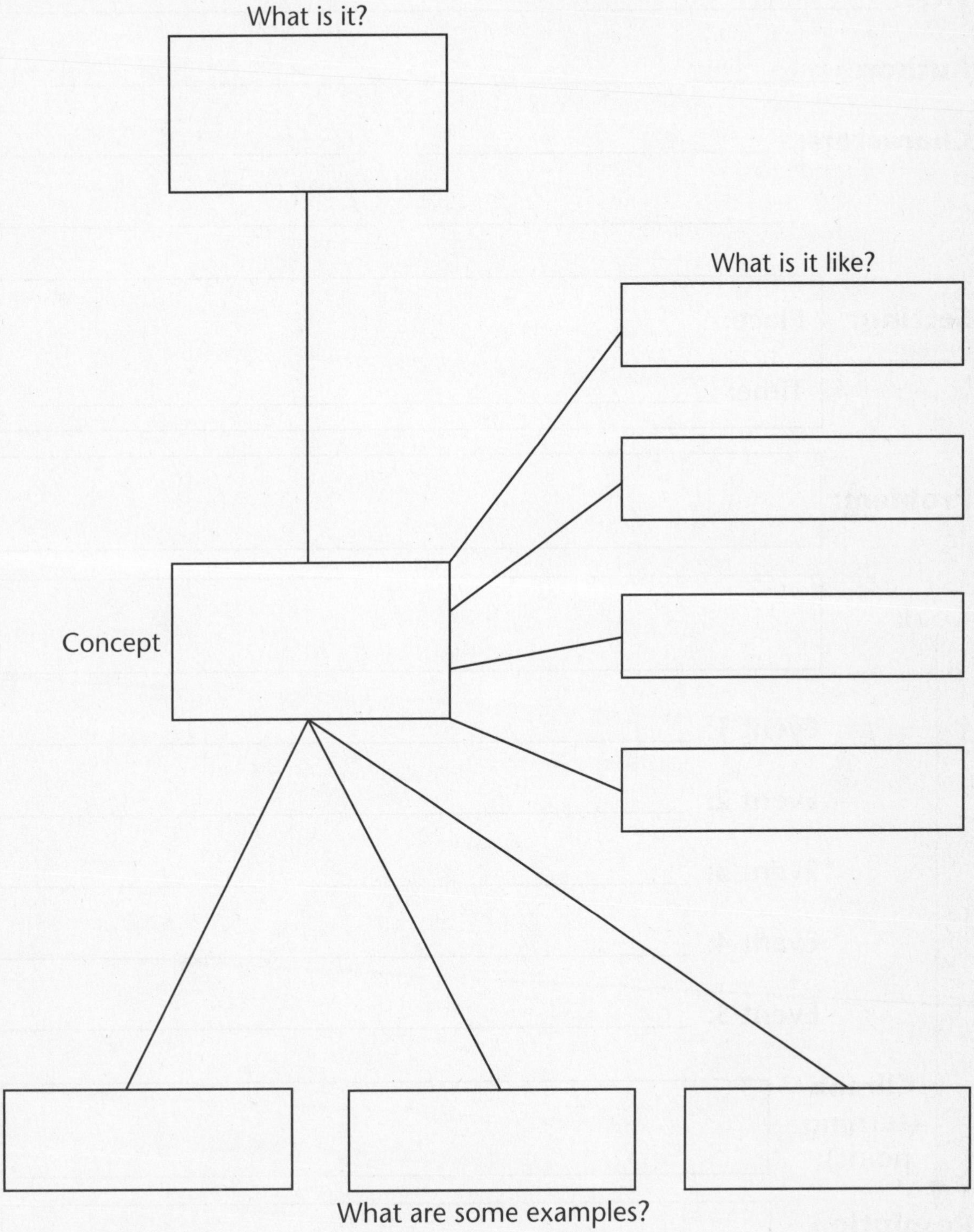

Name:________________________________ Date:________________

Story Map

Title: ________________________________

Author: ________________________________

Characters: ________________ ________________

________________ ________________

Setting: Place:

Time: ________________________________

Problem:

Goal:

Event 1: ________________________________

Event 2: ________________________________

Event 3: ________________________________

Event 4: ________________________________

Event 5: ________________________________

Climax (turning point):

Resolution (conclusion):

Name: ______________________ Date: ______________

Essay Planning Sheet

Introductory paragraph

Lead: ______________________

____

____

Thesis statement: ______________________

____

Body paragraph 1

Topic sentence: ______________________

____

Detail sentence: ______________________

____

Detail sentence: ______________________

____

Detail sentence: ______________________

____

Body paragraph 2

Topic sentence: ______________________

____

Detail sentence: ______________________

____

Detail sentence: ______________________

____

Detail sentence: ______________________

____

Body paragraph 3

Topic sentence: ______________________

____

Detail sentence: ______________________

____

Detail sentence: ______________________

____

Detail sentence: ______________________

____

Concluding paragraph

Restatement of thesis: ______________________

____

Conclusion: ______________________

____

____